CHENKUMBI LIME

Fuel-efficient high-quality production

John Spiropoulos

Practical
ACTION
PUBLISHING

INTERMEDIATE TECHNOLOGY PUBLICATIONS 1992

Practical Action Publishing Ltd
25 Albert Street, Rugby,
Warwickshire, CV21 2SD, UK
www.practicalactionpublishing.com

First published in 1992
Transferred to digital printing in 2008

A catalogue record for this book is available from the British Library & Library of Congress

ISBN 978-1-85339-144-6 Paperback
ISBN 978-1-78044-397-3 Digital book

Citation: Spiropoulos, J. (1992) *Chenkumbi Lime: Fuel-efficient high quality production*, Rugby, UK: Practical Action Publishing https://doi.org/10.3362/9781780443973

Since 1974, Practical Action Publishing has published and disseminated books and information in support of international development work throughout the world. All print editions are produced and distributed via ethical and sustainable print on demand global facilities.

Practical Action Publishing is a trading name of Practical Action Publishing Ltd (Company Reg. No. 01159018 | VAT 880 9924 76). All profits are covenanted back to its parent group, Practical Action (Charity Reg. No. 247257).

Cover photo: Charging limestone into a vertical shaft kiln (Practical Action/Paul Harris)
Design and artwork by DesignWrite Productions, Rugby

The manufacturer's authorised representative in the EU for product safety is Lightning Source France, 1 Av. Johannes Gutenberg, 78310 Maurepas, France. compliance@lightningsource.fr

CONTENTS

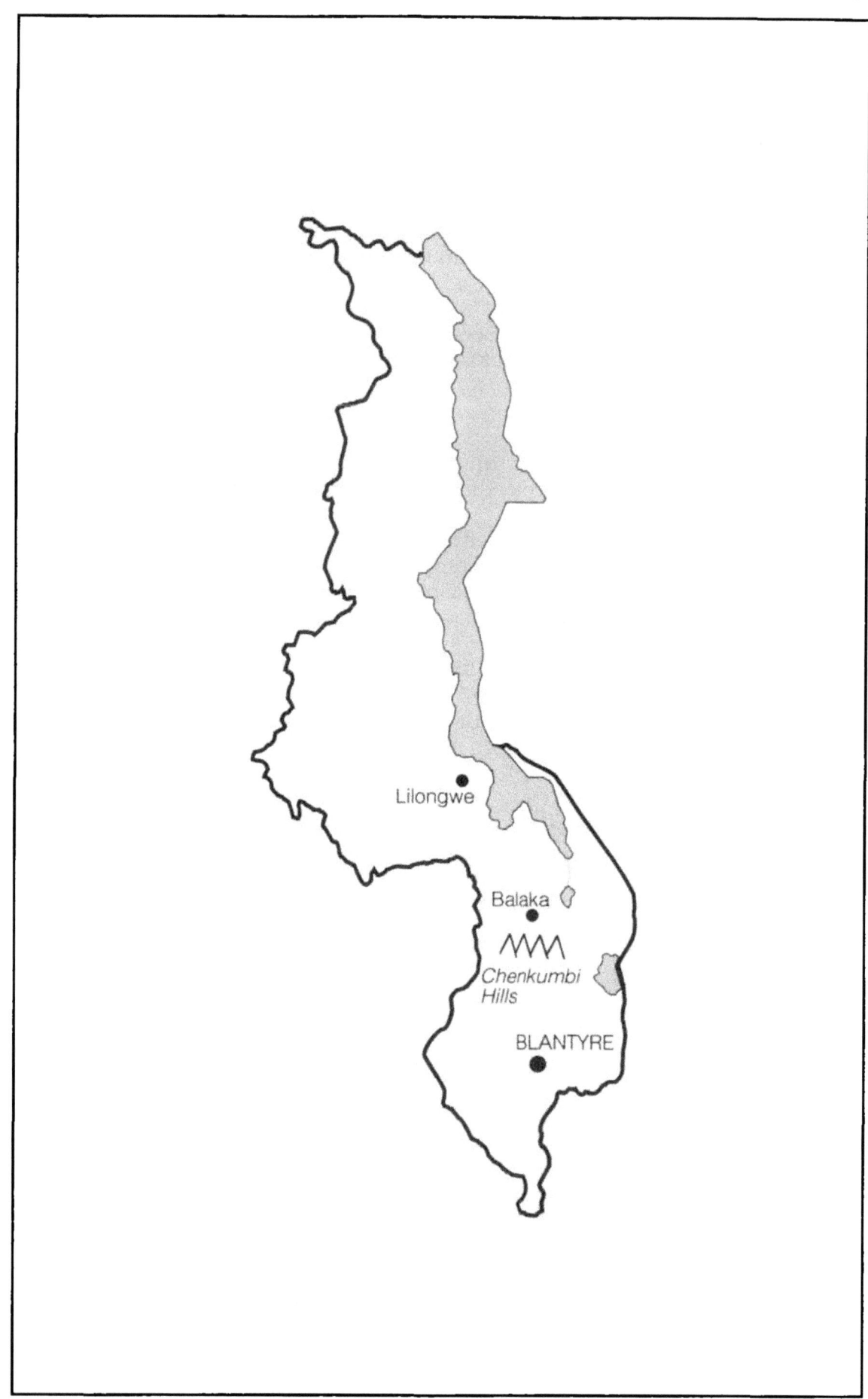

Figure 1. Map of Malawi

INTRODUCTION
The case for lime

The Intermediate Technology Development Group (ITDG) is particularly interested in lime as an alternative binder to Ordinary Portland Cement (OPC) and the case for lime as an alternative binder is given below. In making that case, we must never forget that lime is not only a building material but an essential ingredient to many industrial processes. The demand for high quality, chemical grade lime in Malawi, for instance, stems from the requirements of the sugar industry. Lime is also used in agriculture and food processing; it is essential in the operation of water treatment plants. One could say it is a pre-requisite for development as well as being a potential source of income for small-scale producers.

The manufacture of OPC requires relatively sophisticated technology, equipment and expertise, and in theory benefits considerably from economies of scale. This has led to a general trend to establish large centralised cement manufacturing plants. The potentially low manufacturing costs of large plants will only accrue, however, if the plant operates at a high capacity. In the poorer developing countries, delays in importing fuels and spare parts mean this is often not the case, resulting in high, rather than low, manufacturing costs. Even where low production costs can be achieved, large centralized plants inevitably result in high transportation and distribution costs.

Consequently OPC is often a scarce and expensive commodity in many developing countries, particularly in rural areas. Furthermore, the equipment, expertise and high grade fuel required in cement manufacture are often imported, using scare reserves of foreign exchange.

In comparison, the manufacture and use of alternative cements have numerous advantages. These can be summarized as follows:

(a) **Economic:** much lower foreign exchange expenditure in both capital and operating costs; decentralized production resulting in considerably lower transportation costs; lower unit production costs.

(b) **Technical:** lower temperatures required during the manufacturing process allowing less sophisticated and less expensive production technology; lower strengths, slower setting times and a high degree of workability are ideal for many basic construction applications.

(c) **Social:** labour-intensive production techniques improve employment prospects for the local population; lower building costs results in affordable quality housing for poor communities; decentralized production gives local responsibility for community-based production and supply; wider availability of cementitious materials.

In order to promote the manufacture and use of lime further, ITDG operates an advisory service, the Cementitious Binders Advisory Service (CAS), which is part of a data bank and information service on building and building materials known as BASIN (Building Advisory Service and Information Network) and operated jointly by four European appropriate technology institutions. Further information may be obtained direct from ITDG.

This case study is one part of a series of publications by CAS intended to inform and influence decision makers, project practitioners and individual entrepreneurs. It is technical in content and assumes a knowledge of the subject although some general information is given in the first chapter. For those starting off on a similar journey, the author, John Spiropolous, demonstrates that with determination, a sense of teamwork and co-operation all things are possible.

Background

In 1984, ITDG was requested by the Government of Malawi to assist in a programme of assistance to their indigenous lime manufacturing industry. The industry had recently re-located to a new district and was experiencing difficulties with fuelwood supply. ITDG carried out several studies in co-operation with both the Malawian government and the artisanal lime burners based in the Chenkumbi Hills area of Machinga district, close to Balaka township in central Malawi. The salient points arising from these studies were:

○ Production of lime is seasonal, with up to 1500 families gaining irregular income from lime burning activities.

○ Production of lime is organized by some 40 artisanal entrepreneurs, of whom four to five dominate, by virtue of their ownership of transport and grinding facilities.

○ The 40 artisanal lime burners are organized formally into an association which has a history of successful lobbying for its own interests.

○ Whilst outwardly appearing to be highly independent operators, the lime burners have, in fact, intimate informal links with each other formed around grinding, bagging and transport operations.

○ The product produced in Chenkumbi Hills cannot meet the quality required for over 50 per cent of existing local demand. Indeed, as the levels of imported, high quality, lime increases the Chenkumbi Hills producers become ever more marginalized by their poor quality product.

○ The traditional burning technology is neither economically nor environmentally sustainable with indigenous hardwood fuelwood being illegally transported greater and greater distances.

In summary, about 1,500 families depend, in part, on lime production for their income. This income is seriously threatened not only by imported, higher quality products, but by the fundamental problem of fuelwood supply. Solving this problem was and remains necessary for sustainable development of the lime industry in the Chenkumbi Hills.

Research and development

An analysis of the studies carried out indicates very clearly that the existing technology is not capable of adaptation to increase its short term sustainability. The initial challenge was, therefore, a technical one of enabling the use of environmentally sustainable fuels. A feasibility study based on the use of coal or plantation softwood fuelwood charcoal indicated that economic sustainability would revolve around an increase in fuel efficiency and product quality.

Conventional wisdom states that the coarse crystalline nature of the Chenkumbi marbles makes it unsuitable for low-tech vertical shaft kiln technology. Such wisdom indicates that hi-tech rotary kilns are the only solution. This case study records the research and development phase of a project which eventually enabled the Chenkumbi marbles to be burnt in low-tech, vertical shaft kilns.

Having achieved this, the challenge moved on to the production of high value added products both to meet existing market demand and secure the economic viability of an enterprise. The study briefly describes an experimental processing plant which was developed and is now producing a high quality product meeting the most stringent specifications.

Dissemination

All the technical aims of the project have now been achieved. The technologies have been developed and proven in working conditions as close to those of the artisanal producers as possible. All the materials required for both the vertical shaft kiln and processing equipment were resourced locally in Malawi and installed by Malawian contractors. Already, some of the more advantaged lime burners are planning and installing their own vertical shaft kilns totally independent from ITDG technical assistance. In recording this success however, we would not wish to give the impression that technical achievement and equitable development are one and the same thing. The vast majority of the artisanal lime burners do not have access to the new technology and are becoming increasingly marginalized by the high cost of fuelwood and competing high quality products. ITDG is continuing to work with those producers to facilitate their access to, and control over, the new technology.

Achievements

Considerable achievements were made both in fuel efficiency and end-product quality. The fuel requirement of the new technology is met from managed plantation softwoods or coal (not indigenous hardwoods) and the operation is sufficiently profitable to sustain the additional transport costs. The quality of the end-product is far superior to anything previously achieved in Malawi and meets the most stringent standards for chemical grade lime. Refractory bricks are now locally produced not only for the lime industry but for other industries as well. Local contractors have the capacity to design and build vertical shaft kilns and the associated processing equipment. This impressive list of achievements was only made possible with co-operation of many individuals and organizations. Indeed, the sense of teamwork and co-operation was, and still is, a hallmark of this project.

David Mather
ITDG

1. RAW MATERIALS, APPLICATIONS AND PRODUCTION

Limestone, lime and lime hydrate, are the most versatile and widely used raw materials of industry. Limestone is a broad term for a widely occurring, abundant sedimentary rock with a range of chemical composition and physical character, including density, hardness, porosity and crystallinity. It is predominantly calcium carbonate ($CaCO_3$) with varying amounts of magnesium carbonate, silica, alumina and iron oxide.

Lime is a product of limestone obtained after calcination (firing). Lime hydrate is produced from lime after the addition of water (hydration).

Raw materials

The quality of limestone most suitable for lime production varies considerably. However, in general, it should have as high a calcium carbonate content as possible. Most applications require an available lime content of over 65 per cent, out of a maximum possible of 69 to 70 per cent. Building lime (lime hydrate) may have a lower available lime content of around 58 to 60 per cent, and the agricultural lime may be anything between limestone and chemical grade lime depending on the requirements of the market.

In principle, a valuable limestone will be one which has a high calcium carbonate content. It should also be hard and strong. This type of limestone will satisfy a wide range of possible applications.

Ultra-high calcium limestone	>97% $CaCO_3$
Calcitic limestone	>95% $CaCO_3$
Dolomitic limestone	5-40% $MgCO_3$
Dolomite	40% $MgCO_3$

However, each industrial application requires and makes use of a specific set of chemical and physical properties, which might not include high chemical purity. The suitability of a limestone also depends on its behaviour in processing. For example, a coarse-grained (large crystal) marble, while perhaps being chemically pure, will tend to disintegrate or decrepitate on firing. Furthermore, commercial importance of a limestone deposit is not only dependent on quality but also on the consistency of quality, and availability in sufficient quantity deposited in a form that allows for economic quarrying.

Evaluation of the suitability of a limestone deposit, therefore, must involve a study of geological features in addition to the physical, chemical and mineralogical properties of the limestone. The geological features will determine the method of quarrying that must be employed and the extent and homogeneity of the deposit. Ultimately these features affect the technical and financial viability of extraction. It should be emphasized that the study of the geological features and their implication must be conducted in the light of the method of production and scale of investment intended, and of the market that is to be supplied. This is important because without this information it is quite easy to judge a deposit unsuitable incorrectly.

Some of the geological features which would need to be looked at include: total thickness of deposit and bedding; overburden ratio; extent of strata and variations in thickness; position of water table; areas of dolomitization; faults and mineralization in and adjacent to faults; cavities and interbedding holding clays and silts; and more.

Although all geological services must be professional, the extent and detail of evaluation must be tailored to the scale of investment planned and the quality requirements of the consuming industry. Clearly, a small investment for a three tonne per day operation may not justify a full geological appraisal. Sufficient investigation is required, however, to ensure that there is enough limestone of adequate quality for the duration of the project.

Applications

Limestone is used in a wide variety of applications (see Appendix 1) including the chemical and metallurgical industries; in manufacturing; in the civil engineering and building industries; in agriculture; in Portland cement production; and in the production of lime. The major uses, by far, are in Portland cement and lime production.

Metallurgical applications
The metallurgical applications are in the ferrous and non-ferrous metal processing industries. The main requirement is chemical purity. High calcium limestone with a minimum of 95 per cent calcium carbonate is used. It is crushed, milled and sized for use in particle sizes of less than 3mm.

Chemical industry
Limestone used in glass and chemical production, acid neutralization and Portland cement production must have high and/or consistent chemical purity. Further, these production processes require absence of organic matter, physical integrity in processing, as well as a precise and narrow particle size distribution. Processing and process control, from the quarry to the bag, requires careful extraction and blending to ensure high and consistent quality, crushing and milling, and accurate sizing.

Manufacturing
The quality requirements of the manufacturing industry are generally concentrated on the physical properties of the limestone. They include particle size distribution, hardness, colour, brightness, oil absorption capacity, surface area and grit content. Some do, however, also require high chemical purity. The applications in this category are:

(a) fillers and extenders in plastic, rubber, paint, paper and putty manufacture
(b) mild abrasive in scourers
(c) fungicides and insecticides, as a carrier
(d) glazes and enamels
(e) flue gas desulphurization and acid neutralization.

The processing requirements are blending, crushing, milling and precise sizing.

Civil and construction industry
This industry uses limestone as: dimension stone (marble) which is quarried in large blocks, cut into slabs and polished; as building stone and 'rip rap' of 200-300 mm diameter; and as aggregate for use in concrete, railway ballast and road stone. It is also used in the form of granules in terrazzo work and as a fine powder filler in asphalt.

The important properties are physical. They include appearance, compressive strength, porosity and abrasion resistance. The processing requirement is crushing and mechanical or manual sizing.

Agriculture

Another important application of limestone is in agriculture. It is used to add magnesium and calcium to the soil and to neutralize acidic soils to improve yields. Chemical composition is an important factor together with particle size and size distribution. Size and size distribution are important as they determine the rate of release of calcium and magnesium oxides into the soil. Limestone is also crushed, sized and used in animal feeds.

Lime and lime hydrate

A major use of limestone is in the production of lime and lime hydrate. These materials, similarly, are used in a broad range of chemical, metallurgical and manufacturing applications. They are also used in construction and the civil engineering industry, and in agriculture (see Appendix 1).

The production of lime and lime hydrate involves extraction and possibly washing and sizing of the limestone kiln feed. The kiln feed is then fired in a kiln to produce lime which is slaked by the addition of water to produce lime hydrate.

The firing of the limestone is the most complex and important part of the production process. It is highly sensitive to a number of inter-related factors. These are the quality of the limestone used, the kiln design and method of operation, and the quality and properties required of the product.

The raw material available for use is very often fixed, to the extent that it is the only material sufficiently close to the market. Its properties, however, have a direct bearing on the kiln design and method of operation that can be employed and the quality of lime that can be produced. It is, therefore, important that the properties of limestones, as they relate to calcination and the product that can be produced, are understood.

Theory of calcination and factors affecting kiln design and operation

Temperature and rate of calcination

Limestone is heated in the kiln to a temperature at which carbon dioxide is driven off. This is dependent on the pressure and CO_2 concentration inside the kiln. Dissociation begins at around 900°C for calcitic limestones and around 725°C for magnesian limestones and dolomite.

CO_2 dissociation begins at the surface of a limestone lump and penetrates inwards to the core. Generally, at fixed temperature and kiln atmosphere conditions, the larger the diameter of the kiln feed the longer the duration of calcination required to complete dissociation, and *vice versa*. There is, therefore, an inverse relationship between temperature and the length of time the limestone is left at that temperature. In practice, the choice of these is a function of kiln feed size, porosity and purity, and the quality and rate of output required.

There is a tendency in practice to reduce the duration of firing to increase the kiln throughput. This can only be done, without seriously affecting quality and kiln efficiency, by increasing temperature. Beyond 1300°C for calcitic limestones and 1100°C for dolomites, however, the quicklime becomes hard burnt and difficult to slake. Generally, a temperature of about 1100° for high calcium limestones and 950° for dolomitic limestones produces the best quality product at a reasonably economic rate of production. Rapid changes in temperature must be avoided to maximize the quality of the quicklime produced. The feed should be pre-heated gradually to the firing temperature.

Kiln feed size
In principle, the kiln shaft should be as long as possible to facilitate a gradual increase of the kiln charge to the dissociation temperature and to maximize heat recuperation from the rising hot gases. The size of the kiln feed that can be used is dependent on its ability to withstand the vertical load in the kiln. It must also avoid abrasion resulting in excess fines which would impede even flow of hot gases through the kiln mass.

The size of limestone feed is also dependent on its porosity. The more porous the limestone the faster the rate of calcination. The fine-grained limestones are generally more porous than the coarse. Therefore, for the coarse crystalline limestones or marble to calcine at the same rate as the fine they must be smaller in size. The mean kiln feed size for porous limestone can be between 100 and 200mm diameter whereas a more dense, medium to coarse grain rock will have to be between 75 and 150mm.

Shape
Kiln feed must be as spherical in shape as possible so that it can calcine evenly. Further, vertical shaft kilns must have ample space (30 to 35 per cent) between the limestone lumps to enable an even flow and distribution of kiln gases through the limestone mass.

A variety of sizes and shapes of kiln feed will result in uneven firing through the formation of channels (chimneys) in the kiln mass through which the kiln gases will flow. Only the limestone immediately around the channel will be calcined. The maximum recommended kiln feed size for a hard, dense rock is 100mm with a size gradation 1:2, i.e. 50 to 100mm range with a mean around 75mm.

Impurities
Impurities in the lime are derived primarily from the lime itself but also from the fuel used in calcination. At the firing temperatures, impurities combine chemically with CaO and MgO to form calcium and magnesium silicates, aluminates and ferrites. The presence of these compounds in calcining limestones impedes the flow of CO_2 by clogging the micropore in the stone.

Chemically, the presence of impurities also reduces the quantity of available lime, e.g., if there are 10 per cent impurities in the limestone there will be roughly 20 per cent in the quicklime produced.

Impurities from the fuels occur when mixed feed firing, especially with coals which have a high ash content. Silica, iron and alumina in ash contaminate both in a free form and in the formation of silicate and aluminates with the quicklime.

CO_2 pressure and concentration
A dense limestone will require a higher temperature to calcine. CO_2 will encounter greater difficulty in reaching the surface of the lump. Similarly, greater pressure will be required to release CO_2 from a larger stone than a smaller one of the same type.

Also, if the concentration of CO_2 in the kiln is high, especially around the surface of the limestone lumps, dissociation is impeded. It is considered necessary for sufficient draft to exist through the kiln to keep the CO_2 concentration low and to clear the surface of the kiln feed of the CO_2 envelope that forms.

Decrepitation and spalling
Decrepitation is the disintegration or fracture of limestone when heated to the calcination temperature. The heat causes an expansion of individual crystals which results in high internal stresses and disintegration. This phenomenon is exclusive to coarse crystalline marbles.

Dolomite is especially difficult to fire. The temperature to dissociate the CO_3 from $MgCO_3$ is substantially lower than for $CaCO_3$. Therefore, when firing at the temperature to calcine limestone, the magnesium carbonate overburns. This results in slow slaking of $MgCO_3$ with consequent problems, specifically when used as building lime.

Kiln technology and choice

Kiln designs and methods of operation vary as follows:
- in the way they use fuel to calcine limestone;
- in the manner in which materials are handled, charged and discharged from the kiln; and
- in the type and way materials are put together in the construction of the kiln.

Traditionally, lime kilns have been, and in many places continue to be, simple, rudimentary structures which fire limestone in batches. These are either flare kilns which incorporate fireplaces in the kiln structure or mixed-feed kilns in which limestone and fuel are stacked in layers.

The advantages of these types of kilns are: very low level of investment, use of locally available materials for construction, intermittent production to match demand for lime and the time available to producers. The characteristic disadvantages of these types of kilns are: very inefficient use of fuel, low quality output, low-level and intermittent production, and high wastage.

However, while there are real disadvantages in using batch-type kilns they might be the most appropriate choice in circumstances where the demand is low and intermittent, and where labour availability is seasonal. Where necessary, it is possible to retain batch production and minimize disadvantages such as high fuel consumption and quality, and high waste, by modifying the kiln design. Traditionally this design modification has been the vertical shaft kiln.

These modifications, however, have not been sufficient to meet the demands of modern industry. Lime kiln designs have have therefore been developed to meet these demands of high quality, low fuel consumption and waste of kiln feed, and a low capital cost per unit labour and output. There are now a large number of kiln types available using different fuels in a variety of ways.

Modern kiln designs adopt a continuous mode of operation which is more fuel efficient and more productive. One type of modern kiln is the vertical shaft kiln of which there are two kinds — the mixed feed and the externally fired.

Mixed feed vertical shaft kilns

There is a wide range of kilns in this category from the most simple to the most sophisticated and fuel efficient. Their major advantages are that they can have a simple design and be easily constructed, they can be flexible in the materials that can be used for construction, and can be operated relatively easily. Further, a full range of solid fuels can be used, including different quality coals, coke, charcoal, wood and many agricultural wastes such as nutshells and oil pips.

Continuously operated vertical shaft kilns work on the basis of the limestone mass passing down the kiln while hot gases flow up the shaft and into the atmosphere. The flow of hot gases up the shaft through the limestone bed implies the presence of a draught. This may be natural due to convection. Alternatively, it may be a mechanically induced draught from the top of the shaft or forced through the limestone mass from the base by electrically driven fans.

Mechanically induced or forced draught systems have the advantage of providing greater control over the calcination process providing, in principle, a better quality product and higher productivity. Also, adoption of induced or forced draught makes it possible to recirculate exhaust gases and create a more fuel efficient system.

Externally fired kilns

There are three types of externally fired kilns: furnace fired, producer gas fired and oil fired. These have been developed to avoid contamination by fuel ash or, in the case of fuel oil and gas, to facilitate the use of a liquid fuel.

Furnace-fired kilns have separate fire boxes which are constructed in the lower middle third of the vertical shaft and in which solid fuels are fired. The hot gases from the fires are drawn into the kiln either by natural or induced draught. The use of wood fuel is preferred in furnace-fired kilns as this provides a long, cool flame which can penetrate more easily into the limestone mass. The disadvantage of such kilns is low fuel efficiency mainly because the fuel is fired outside the limestone mass.

Producer gas fired kilns are where the firebox concept of furnace fired kilns was modified to improve fuel efficiency. They were modified into gasifiers which convert solid fuel into producer gas which is drawn into the kiln and ignited inside it. This system can also burn fuels which would otherwise be very difficult to use. For example, lignite, peat and highly volatile coal and certain agricultural wastes.

Oil-fired kilns, in their simplest form, utilize heavy fuel oil (bunker c) which is atomized by steam or air under pressure and sprayed onto the hot limestone mass in the kiln from combustion chambers set into the wall of the kiln.

The more sophisticated kilns gasify the oil in sealed gasifiers. The hot gases produced are then drawn into the kiln mass where they are ignited. These kilns are generally more fuel-efficient but they are more complicated in design, and difficult to maintain and operate. The primary advantage of oil-fired kilns is that it is possible to produce a uniform and consistent quality lime with very little contamination. Unfortunately, however, in developing countries they are generally feasible only in those countries which produce their own oil. There are also shaft kilns in the upper end of the 1 to 100tpd range which are designed to use natural.gas.

Rotary kilns

Despite the higher fuel consumption compared with modern vertical shaft kilns, rotary kilns continue to be used, especially in the US, where fuel is relatively cheap and labour expensive. The advantage of rotary kilns is that they can use a smaller kiln feed size of 10 to 50mm diameter. Kiln capacities range from 25 to 600tpd, although in practice it is very difficult to find kilns with a capacity of less than 100 tonnes per day.

Most lime calcination technological developments have happened in the context of the economic, technical and climatic environment of northern hemisphere industrialized countries. This has often resulted in technology which is highly effective but inappropriate for many southern hemisphere developing countries. Even in instances where northern kiln suppliers have scaled down versions of their kiln technology to meet the requirements of developing countries for small-scale, decentralized production, the technology often remains unsuitable.

However, more and more new kiln designs, often adaptations of sophisticated kilns from industrialized countries, are being developed and tested in the developing countries of the South to solve particular local problems. The case study in this publication, describing a low-cost kiln technology development in Malawi, is one such example.

2. BACKGROUND

Project location and environment

The Chenkumbi Hills is located 15km south of Balaka town on the main road between Lilongwe and Blantyre. It has a semi-arid climate with rainfall too erratic for successful agriculture. Soil cover is thin and the area has been practically denuded of trees which have been used intensively since 1985 by the lime producers.

Social and economic survey

A social and economic survey was conducted to inform the project planners of the context in which the project was to be implemented. Furthermore, the survey was intended to identify and characterize the target group to enable the project to be designed for its maximum benefit.

The Chenkumbi Hills area has a population of around 2500 and an estimated population growth rate well above the national average. Agriculture has already started on marginal land because of over-population in the district and the shortage of suitable land.

Land use and population surveys in the last census have showed that 48 per cent of the total population of the district is supported on land holdings of less than one hectare. Farms of this size produce less than half of family nutritional requirements, so income from off-farm activities is crucial for the survival of a large section of the resident population. A survey of the population working in the lime industry showed that all the workers fell into this category. This information showed clearly the need to maintain or even increase the current level of income flow to the community.

The majority of traditional lime producers operate seasonally producing only one or two batches per year, between March and December. A few who operate on a more commercial basis produce between three and five batches per year. Each batch averages approximately 50 tonnes of product.

Employment is seasonal, casual and insecure. Typically, wages are well below the government minimum and workers are paid when owners' cashflows permit.

Mineral resources

The limestone resources of the Chenkumbi Hills area are the largest single known deposit in Malawi. The measured reserves of the main limestone group stands at 3.7 million tonnes which is more than adequate for the foreseeable future. The reserves are a coarse grained calcitic marble of a sufficiently high chemical purity for the production of chemical grade lime.

Typical chemical analysis of Chenkumbi limestone (%)	
Calcium Oxide (CaO)	52.06
Magnesium Oxide (MgO)	2.23
Inerts and traces	2.69
Loss on Ignition	43.02
TOTAL	**100.00**
CaO content of limestone	*91.4*
Available lime content	*69.7*

Fuel resources

A fuel resources study was carried out to provide information on fuel supply, existing reserves, current consumption and projected demand. Initially, it was concluded that existing government plantation woodfuel reserves would be available for lime production. However, with the increasing influx of Mozambican refugees in the vicinity of the project area, the need for plantation fuelwood for the domestic use of this population has resulted in a change of policy by government.

It was realized that both plantation and indigenous forest reserves in the area were severely limited and that alternatives sources of fuel would need to be used in future. Eventually, it was recommended that the lime producers plant their own plantations, or they would have to use the commercial charcoal from Vipya plantation or the coal from the Northern region.

Market for lime

A market study showed the market to be composed of two parts. The first is for high-grade lime used predominantly by the sugar industry and the second is for a lower quality product consumed by the construction industry and to a certain extent by the agricultural sector.

The sugar industry consumes about 3200 tonnes of lime per annum imported mainly from Zambia and South Africa. The demand for high quality imported lime is fairly stable and is concentrated between April and November each year. It is estimated that the demand for this quality will increase at a rate of about five per cent per annum over the next 10 years.

The quality specifications demanded by the refinery are high but, in practice, the quality imported has often been variable and well within the reach of local production (see Table above). The refinery has, from time to time, bought from local producers and blended their product with higher quality imported lime.

The demand by the construction industry is less stable. It is subject to the usual fluctuations of the industry and also seasonal variation in construction activity which is concentrated in the dry season.

At this stage of the project, it became apparent that a technology needed to be developed and introduced to the local producers which would open up the high grade market to them. In addition, it would have to produce a consistent quality and quantity to enable stabilization and an increase in demand by the construction industry.

Preliminary investigations

Investigations were initiated with a rapid appraisal of the performance and methods of operation of the traditional producers together with a desk study of the work done by the geological survey department. This showed clearly that kiln design was the major obstacle to the production of higher quality lime and more fuel-efficient production.

A series of tests were then carried out on the limestone, both in the laboratory and in the field, to determine specifically its firing characteristics. Unfortunately, the tests showed that a simple solution would be unlikely. The limestone available is a particularly difficult rock type. Although it is chemically of a high quality, it is a coarse crystalline marble which is difficult to calcine because of its friability on firing.

The effect is fragmentation of the limestone/lime in the kiln and restriction of the airflow which is necessary for adequate calcination. Calcination is thus incomplete and the quality is low. Furthermore, because of the restricted air flow through the kiln mass, there is very little heat recuperation or cooling of the lime mass. This results not only in a substantial waste of energy, but the lime also becomes very difficult to handle.

ITDG consulted a number of specialists in the field and it was concluded that it would be possible to improve the quality with a suitably designed and operated kiln.

3. TRADITIONAL PRODUCTION

A detailed physical measurement of a typical kiln batch was made, from the quarrying of the limestone to the bagging of the lime hydrate. The reason for this was to develop an insight into the technology employed, the methods of operation and work organization of the traditional producers, and in this way to have a basis of comparison with any new method of production.

Production is concentrated in the period from March to December with only one or two producers operating throughout the year, albeit at a much reduced rate. This is primarily due to the impassable roads, especially into the forest reserves from which fuelwood is collected.

The typical form of organization is a single owner performing the management function, a small core of permanent employees who do the more skilled and supervisory tasks and casual labour used wherever possible and paid on a piece-rate basis.

Production cycle and technology

The production cycle begins with manual quarrying of limestone. Picks, hammers and crowbars are used to extract pieces of surface limestone of between 200 and 300mm diameter. These are then stacked one metre high and 500mm in diameter, making a 'heap' which constitutes a day's work by one person.

The limestone pieces are then broken further with 2.5kg hammers to kiln feed size of minus 50mm in diameter. Dressing of one standard heap constitutes a day's work.

A typical kiln batch requires about 75 tonnes of limestone and, in the operation observed, it took 15 days to extract and dress the required limestone kiln feed.

Traditionally, producers use locally available indigenous hardwood. The kiln technology used does not allow for the use of fast growing plantation fuelwoods.

Figure 2. A traditional lime kiln in the Chenkumbi Hills.

Measurement of fuel use showed that between 50 and 55 tonnes of wood are used per kiln batch.

Kiln loading commences when both the limestone feed and the fuelwood are on site. There are, however, sometimes long delays between kiln batches because of difficulties in obtaining fuelwood.

The kiln is a rectangular box-type constructed of limestone boulders cemented with lime-mud mortar. The kiln walls are buttressed. Sometimes one long side of the kiln is built lower than the other and buttressed with sand for access to charge the kiln. It has two firing openings at the base on the two short sides each leading into two trenches running along the length of the kiln (see Figure 2).

Arched vaults are built over the trenches with large limestone boulders and the kiln is charged with five alternating layers of fuel and limestone, starting with kindling at the base and limestone feed at the top.

The kiln is ignited and then stoked for about 48 hours. It is then sealed and left to burn out and cool before discharging begins. The kiln observed took eight days to burn out and cool.

The kiln is discharged from the top. Often this is done in two or three batches, partly due to the lack of space available for storage of lime but, more importantly, because most producers only carry out slaking and milling once a sale is secured.

The lime discharged from the kiln is mostly fragmented to minus 5mm in diameter. This is due both to the friable character of the rock and air slaking.

When the quicklime is finally slaked it is done by pouring water over a large pile which is then turned manually with a spade in the same way that concrete is mixed. Due to air slaking and also due to the method of slaking itself, the material produced is often not thoroughly hydrated.

The hydrated lime is then usually sieved through a hand punched sheet metal sieve to remove the coarse unburnt limestone. In practice limestone of minus 10mm passes through the sieve and is taken to milling with the finer lime hydrate. Due to the crude sieving, a quantity of unburnt limestone cores is milled together with the lime hydrate which lowers the quality of the final product. This is done in locally produced hammer mills, usually supplied and used for maize milling. There were three such mills as well as a small mineral-processing hammer mill serving between 17 and 22 producers in the Chenkumbi hills area.

The milling is not very efficient as the feed from slaking is milled without any pre-screening of the fines. Milling of the larger particles is thus much less efficient, resulting in the presence of quite coarse material in the final product.

The hammer mills are powered by diesel engines and the hammers (beaters) are usually made of mild steel and sometimes of spring steel. Diesel consumption is high and breakdowns often cause lengthy delays. Although the producers have become quite adept at repairs, bad maintenance of screens and worn beaters results in inefficient production and poor quality product.

The bagging operation is performed manually, sometimes into three-ply paper sacks which have become prohibitively expensive, but most often into second-hand fertilizer or relief food sacks.

Input/Output

Some 70 to 75 tonnes of limestone and 50 to 55 tonnes of fuelwood is charged into the kiln. The loss on ignition is 25 per cent leaving 56 tonnes of quicklime and unburnt core waste of plus 5mm. Eight tonnes of water for hydration is added making 64 tonnes. Twenty to 25 per cent of this is sieved out as plus 5mm material leaving about 50 tonnes (2000 25kg pockets) lime product which goes to milling. The production cycle averages 60 days per batch.

Quality

The typical chemical composition of the Chenkumbi Hills limestone implies that the lime that can be produced by complete calcination of this rock would have a CaO contnet of approximately 91 per cent and a corresponding Available Lime Content (ALC) of 69 per cent.

The quality of the lime produced by the producers using the traditional technology ranges between 32 per cent and 45 per cent ALC. Usually it is between 38 and 40 per cent.

The quality is low due to:

a) poor selection of limestone kiln feed which is derived from the casual limestone extraction method;

b) kiln design which allows only about 50 per cent of the limestone kiln feed to be thoroughly calcined;

c) air slaking and the crude method of screening and hydration;

d) no pre-screening of hammer mill feed, poor maintenance of screens and hammers.

Fuel consumption and efficiency

Measurement of the bulk density of the fuelwood used showed it to be 475kg per m^3, say half tonne per cubic metre. The kiln batch observed used 14 truck loads of about $8m^3$. This means that about 55 tonnes of indigenous fuelwood is used per kiln batch or about 2000 tonnes per annum by the whole industry.

The thermal efficiency of the kiln is calculated to be just below 15 per cent. What this means is that only 15 per cent of the fuel consumed is used to calcine the limestone. The remainder (85 per cent) is lost to the atmosphere.

Project implementation strategy

The problems faced by the small producers are directly related to the technology they use and the method of operation associated with it. Additionally, they do not have sufficient working capital, and there is generally also a low level of business skills. It was concluded that the stabilization of the local inductry would require effort in all areas of weakness. However, it became clear that the development and introduction of a suitable technology was the top priority. Without this the local industry would not be able to supply market need and would become marginalized.

The new technology would have to fulfill a number of criteria but first and foremost it would have to produce lime for the high quality market and thus expand the supply opportunities of the small producers. Additionally, the kiln design would have to reduce fuel consumption dramatically and preferably use an alternative fuel type to the indigenous hardwood. It was also required to at least maintain the current income generation and employment levels within the community. Furthermore, it would have to be compatible with the level of operating and technical skills available and be accessible to the industry.

Finally, the technology would have to use local resources, including materials and engineering capacity, and be sufficiently robust to withstand varying standards of operation and maintenance. Given the difficult quality of the raw material, development of a lime production technology which would meet the design criteria became a daunting task. It was, therefore, decided to proceed in a phased manner. The kiln technology was to be the first part of the process to be developed since there was a certain amount of unceratinty about whether the necessary available lime content could be achieved. This would then be followed by incorporation of a suitable hydration and milling and sizing plant.

The first step in the kiln technology development process was to examine the potential of the simplest option, the natural draft vertical shaft kiln.

4. NATURAL DRAUGHT VERTICAL SHAFT KILN

The first, experimental, vertical shaft kiln constructed at the Chenkumbi Hills was a simple natural draught kiln. It was constructed of fired bricks produced in the area, and limestone boulders. The kiln was built on 400mm thick concrete foundations set on bedrock, with a limestone masonry platform raised to 500mm above ground level. A 150mm concrete slab was then cast on top of this, forming the base for the kiln shaft. The kiln shaft was 4.2m high and had an internal diameter of 1.1m. The kiln wall itself was built with a 400mm thick limestone masonry casing, 330mm fired brick wall internally and a temporary kiln lining of fired brick. Three inspection holes were left in the kiln wall, one in the middle of each third of the kiln, i.e. bottom, middle and top.

Bar rings of 18mm diameter mild steel were to be built into the masonry every 400mm up the shaft to reinforce the kiln and prevent cracking but, unfortunately, due to bad supervision and workmanship they were ommitted. Consequently, with repeated firings, the kiln started to crack. Due to the uneven shape and surface of the kiln it was not easy to install the reinforcing bars or steel straps after the kiln construction was completed. Since the kiln was viewed as experimental, it was decided to begin the firing trials without reinforcing. The kiln was constructed by local builders with little experience in conventional brick and mortar construction, and no experience in kiln construction.

After the first firing trial the kiln was modified by extending its length to 6m and gradually narrowing the internal diameter of the shaft from 1.1m at one metre from the top to 800mm diameter at the top.

The kiln was located at the base of a hill from which limestone of suitable quality was quarried. The quarried limestone was dressed in the quarry area to kiln feed size and charged directly into the top of the kiln via a timber-charging ramp.

Figure 3. Experimental vertical shaft kiln in the Chenkumbi Hills.

Quarrying

The limestone was quarried manually using crowbars and hammers, and dressed with 2.5kg hammers to kiln feed size. Initially, the kiln feed size was set at 100 to 150mm diameter but later this was reduced to 75 to 125mm. Strict quality control was applied to ensure the size limits were adhered to. There were 15 people occupied in quarrying and dressing: seven in quarrying, seven dressing and one supervisor who was also responsible for supervision of firing. They worked only one eight-hour shift per day.

Kiln firing

Firing in the simple verical shaft kiln was done with the fast-growing plantation fuelwood. The objective was to test its usefulness as an alternative to the hardwood used traditionally. In addition, it was intended to show that this type of kiln could be more fuel-efficient and could improve the quality of the lime produced sufficiently to make possible sales to the sugar industry.

The kiln was charged continuously, 24 hours per day. Six people operated the kiln during the day shift (charging and discharging) and four in each of the two night shifts, including the supervisor. The charge rate during the firing trial was 300kg limestone per hour and 120kg fuelwood. The firing temperature reached in the hottest zone, i.e., towards the top of the bottom third of the kiln shaft, was 950 to 1050°C.The temperature at the top inspection hole was between 750 and 850°C, and the temperature at the base, measured at the discharge opening, was 700°C. This indicates that the firing zone in the kiln is between 4 and 4.5m long out of a total kiln height of 6m.

The firing zone was very long. This is caused by the tendency of the rock to decrepitate or become friable on firing. Fragments of kiln feed break off in the kiln due to abrasion of the material as it passes down the shaft and also due to the compressive load of the limestone mass itself. It becomes compact, and air flow through it is restricted. The result is a tendency towards the channelling of air flow, inadequate heat distribution through the limestone mass, poor fuel use and insufficient cooling. Limestone only in the immediate vicinity of the channel is adequately calcined. Also, unburnt charcoal was being discharged from the kiln together with the lime. This is a waste of fuel. The long firing zone also resulted in the discharging of hot lime which was necessary to keep the fire from rising out through the top of the kiln. The temperature of the discharged lime was between 300 and 400°C. This is a waste of heat.

The lime was discharged from the kiln by shovel into wheelbarrows and taken to the slaking shed. After discharging material the base of the kiln shaft was prodded with a metal crowbar or lengths of timber to break any arching which might have occurred in the kiln and to bring the mass down, and draw hot lime out into the discharge ports to cool for the next discharge. The loss on ignition (LOI) of the material discharged was about 27 per cent.

Slaking

The lime was slaked manually on a concrete slaking floor. Small heaps of lime, 30 to 40kg, were separated with a garden fork into two segments, one plus 20mm and the other minus. The two size fractions needed to be slaked separately because they hydrated at different rates. The heaps were flattened out at the top and then water was sprayed on them from a watering can until the material at the surface of the heap appeared to be saturated. The operator waited for a few minutes, working on a second and third heap and then, using the garden fork, the unslaked lime was lifted to the surface of the heap and re-sprayed. This process continued until the material broke down to a slightly moist powder which could be turned easily with a shovel. Approximately 25 litres of water were used to slake 38kg quicklime. Given the quality of lime produced (LOI 27 per cent), 11 litres of water were used to slake the lime and the remainder evaporated. The slightly moist and hot slaked lime was then shovelled to a large heap where it was left to continue the slaking process, dry out and cool for 24 hours. Slaking was done only in the day time shift.

Two teams of three people each were occupied in slaking during the trials. However, it became obvious during the trials that this workforce was not sufficient to cope with the kiln throughput. Also, the slaking floor area was not enough. The area used was 50m². Two slaking sheds of 60m² each are required, with two teams of three in each shed.

Screening, milling and bagging

This part of the process was carried out in the traditional manner. The lime hydrate was sieved through a 5mm sieve, taken to milling in the traditional hammer mill, and bagged manually. The workforce required to do the sieving, milling and bagging was the same as for the traditional method.

Input/output

Approximately nine tonnes of limestone were quarried to produce 7.2 tonnes of kiln feed. This constitutes about 30 per cent waste in dressing. The daily limestone kiln feed was 7.2 tonnes and the fuel consumption was 2.88 tonnes. The charging rate was 300kg limestone and 120kg fuelwood per hour.

The discharge rate of quicklime was about 220kg per hour or 5280kg in 24 hours. The quantity of water used in slaking was about 660 litres per tonne of quicklime. Some 150 to 155 of the 660 litres was used in hydration, the remaining portion evaporating during the process. The total amount of water used per day was approximately 790 litres. Nearly 1000kg of limestone was sieved off per day as waste, i.e. plus 5mm material (equivalent to about 4300kg CaO in 24 hours). This left a product for milling of around five tonnes or 200 25kg pockets of lime hydrate per 24 hour day).

Quality

The quality of lime produced from this technology ranged between 42 and 48 per cent available lime out of a possible 69 per cent. Most analyses of the final product during the trials were between 44 and 46 per cent although the quality seemed to improve as the trial progressed and the operators became more familiar with the kiln operation. It was felt that the priority was to raise the available lime content first before approaching the issue of fineness. For this reason a size analysis was not carried out at this stage. However, visual and physical inspection did indicate that the limestone was calcined more effectively. The product contained more fine material than that produced with traditional technology.

Fuel consumption and thermal efficiency

The fuel used was fast-growing plantation eucalyptus. The fuel consumption was 400kg fuelwood per tonne limestone charged into the kiln. This is a ratio of 2.5:1. This is not the maximum fuel efficiency achievable with plantation fuelwood. It should be possible to use a ratio of 3:1 and even 3.2:1. The reason for the relatively high fuel consumption was that the trial was conducted during the rainy season. Also, the fuelwood used was green and had been cut very recently. Ideally the cut fuelwood should be left to dry in the open for at least four months and preferably six before use. Green timber has only half the burning efficiency of dry. Up 50 per cent of the wood's energy content is used to evaporate the moisture in it. The thermal efficiency of the kiln was calculated to be 30 per cent (see Appendix 2). This could be improved to 36 per cent if the fuelwood is air-dried before use.

5. EXPERIMENTAL FORCED AIR KILN—CHENKUMBI

The natural draught kiln with all its improvements, including dramatically reduced fuel consumption, higher productivity and increased profitability, failed to interest the lime producers. An informal survey of opinion to assess the reasons for this indicated that although producers were generally impressed, especially with the ability of the new kiln to use plantation fuelwood, which is more readily available and cheaper than indigenous hardwood, the other advantages of the new kiln were secondary. The dominating goal of the producers was to supply the high quality market and, clearly, the new kiln could not improve the quality sufficiently to do this.

Furthermore, adopting the new kiln constituted a considerable investment for the producers coupled with a perceived high risk. None of the producers had direct experience in operating vertical shaft kilns. They were therefore uncertain as to whether they could achieve the improvements obtained during the trials by ITDG. To the extent that no interest to take up the new kiln emerged, the technology and the strategy adopted can be considered a failure. The experience, however, provided essential information on the behaviour of the rock when fired and gave clear directions for the next steps in research and development.

The main reason for the inadequate quality was poor firing, apparently due to inadequate air flow and heat distribution in the kiln. The main target thus became to improve quality. It had already been established that fuel efficiency could be improved considerably. The criteria for design also became more focused. It was clear that the new kiln had to be constructed with locally available materials and engineering capacity and be robust in use, i.e. not be very sensitive to operator error, but be compatible with the technical and operating skills of the producers.

Forced air technology
It was decided to adopt a forced air kiln system. In this system air is forced into the kiln (through a tuyère) under pressure at the base of the shaft. It was felt that this type of system would provide the necessary measure of control over the firing conditions, specifically the length of the firing zone and the temperature and heat distribution, to improve the quality of lime produced. In addition, the system is relatively inexpensive and robust compared to induced draught systems which draw air through the kiln from the top.

The forced air system used was observed by the author in use in western Kenya and adapted to suit the scale of operation in Malawi. It consists of cast iron units fitting on top of each other to form a manifold. The manifold is built one metre high on top of a 500mm pyramid located in the middle of the floor of an ordinary vertical kiln shaft. Air is introduced into the manifold and the kiln by means of a fan blower through a 150mm diameter pipe.

Chenkumbi Hills — experimental forced air system
At this stage in the life of the project, it was becoming quite clear that the use of plantation fuelwood could not be a long term solution for the industry. Furthermore, the experience with its use in the natural draught kiln, given the

physical quality of the limestone feed, suggested that its propensity to ignite at comparatively low temperatures could have been one of the causes of the long, and unmanageable firing zone.

The first objective of the forced air trials, therefore, became to show that the kiln could operate successfully and efficiently using charcoal and coal as fuels. Also, the government of Malawi was strongly promoting the use of these fuels as an alternative to the plantation fuel. A second, equally important, objective was to prove that the technology could produce a suitable quality lime for the sugar industry.

Raw materials

The limestone used was analysed and shown to have a CaO content of 91.4 per cent which corresponds to an Available Lime Content of 69.7 per cent. The limestone feed size was between 75 to 125 mm.

The fuel used in the trials was a charcoal which was shown to contain:

Fixed carbon	85%
Volatiles	12%
Ash	3%
Total	100%
Net calorific value	30MJ/kg

The charcoal feed size was from 100mm diameter down to sand size.

The kiln

The kiln used for the trials was the natural-draught kiln described in chapter 4, modified to incorporate the forced air system. It was a brick and stone masonry structure with a shaft height of six metres and a diameter of 1.1m.

The top metre of the shaft converged to an opening of 800mm and a 2m high sheet metal chimney structure, incorporating a hood and a door for charging, was made and installed over the top of the kiln. The main purpose of this was to

Figure 4. Forced air system initially installed in the Chenkumbi Hills, and then transferred to Balaka.

clear the environment at the top of the kiln of exhaust gases and thus make for healthier working conditions for those charging the kiln.

Three inspection holes had been built into the kiln wall at 1.5m, 3m and 4m from the floor of the kiln. A centrifugal fan blower with a rated capacity of $1,000m^3/hr$ against 20 inches water gauge provided the air input. This was driven by a small diesel engine. The blower was overdesigned. This was shown clearly during the trials and also during commissioning of the first production kiln at Balaka.

During the trials large cracks developed in the kiln, and these were probably forced wider by the stone inside the kiln. The reason for this was omission of adequate reinforcing as described in Chapter 5.

Kiln operation

The variables that determine kiln operating efficiency are:
- ○ kiln design including shaft height, diameter and height to diameter ratio;
- ○ materials use in kiln construction, especially the insulation;
- ○ limestone and fuel type and quality, and kiln feed size and size distribution;
- ○ limestone-to-fuel ratio;
- ○ air flow through the kiln;
- ○ frequency of discharge.

The first three variables were taken as fixed, although the kiln feed size could have been varied.

The method of experimentation was to set the second group of three variables and then run the kiln for sufficient time to allow the operation to stabilize as indicated by relatively constant temperatures at the inspection holes. A sample of burnt lime which had passed through the firing zone at stable conditions was taken, slaked and analysed on site.

The operating conditions that were varied to change the performance of the kiln were as follows:
- ○ **Frequency of discharge** was used to vary the position of the firing zone and also the retention time of the limestone in the kiln. The same amount of lime was discharged from each discharge door at the set discharge rate. Each discharge was between 220 and 240 kg. The charging rate was determined by the rate of discharging. The kiln was fed when a space developed after a discharge.
- ○ **The air flow** determines the rate of combustion of the fuel and, therefore, the length of the firing zone and the temperature. Air flow was controlled by means of a damper at the fan inlet and a variable bleed at the discharge side of the fan. It was measured indirectly by recording the pressure drop across an orifice plate in the ducting connecting the fan to the kiln. A simple manometer fixed to a board consisting of a plastic tube with coloured water in it and a school ruler were used to measure the pressure difference in millimetres. A calibration curve of pressure drop (mm) against air flow was prepared and used.
- ○ **The stone-to-fuel ratio** was varied to change the temperature in order to achieve optimum fuel efficiency and calcination.

In some instances it became apparent that the conditions set were not going to work. In these cases the trial was terminated immediately. For instance, high air flows resulted in a very compressed firing zone. Consequently the retention times were too short. Low air flows, on the other hand, led to incomplete combustion. Rapid discharging led to the firing zone dropping too low in the kiln and heat being lost in the discharged rock. Six trials were carried out over a period of 300 hours. Data from the first trial was incomplete and, therefore, not analysed.

Results

Table 1 summarizes the conditions and results for each trial. Trial 6b was conducted with a 50:50 wood:charcoal mix as fuel. All other trials were fuelled with charcoal only.

Temperature

In conducting the trials it was decided to position the firing zone at the centre of the shaft, and to aim for temperatures of 1000 to $1050°C$. The temperature measuring probes (thermocouples) were not able to penetrate more than 200mm into the shaft beyond which temperatures may have been higher.

Discharge rate

The retention time required at these temperatures for an average kiln feed size of 100mm is calculated to be between eight and 10 hours corresponding to charging rates of between 110 and 220kg/hr. However, reference to Table 1 shows a gradual improvement through the trials, the most noticeable improvement being the result of reducing the discharge rate to 105 minutes. This corresponds to a charging rate of 240kg/hr, which is somewhat higher than expected. The difference is probably due to decrepitation of the rock on firing, resulting in compaction.

The optimal discharge rate for the given fixed variables is somewhere between one hour 30 minutes and one hour 45 minutes.

Air supply

The amount of excess air being put through the kiln mass was much higher than expected, although the measured air flows should be treated with some caution given the difficulties in achieving accuracy. The main reason for this was air channelling its way up the sides of the shaft, which was noticeable especially during the last trial with wood where temperatures at the centre of the top of the kiln were much higher than close to the walls. Smoke (unburnt volatiles) was seen to come predominantly from the centre. Lower excess air rates generally resulted in lower temperatures, although combustion was mostly fairly good. Certainly, no smoke was ever visible, but some charcoal was discharged.

Increased air flow increases the rate of combustion and hence the temperature in the kiln. This occurs up to a point beyond which the air simply cools the limestone mass and kiln structure, resulting in thermal inefficiency. In the final analysis, high excess air rates are not important provided the efficiencies are high.

Input/output

A discharge frequency of one hour 45 minutes produces approximately 140kg output (CaO) per hour. This is 3360kg in 24 hours. The amount of limestone charged into the kiln was about 5.76 tonnes and the amount of waste screened after slaking was about 500kg per day.

The CaO product was, therefore, in the region of 3000kg per day. Approximately 1.4 tonnes of water was used per day and 795kg of this combined with the CaO to produce the lime hydrate. The remainder evaporated in slaking.

Approximately one hundred and fifty 25kg bags of product were produced per day.

Product quality

The decrepitation is inherent to the rock type and is discussed elsewhere. Its effect is the production of two products. The outer surfaces of the rock is thoroughly calcined, breaks off, and forms a high quality product (fines). The inner core of the rock is only partially calcined or remains uncalcined. This is the second, lower quality, product (see Table 1).

Using the manual method of slaking described in previous sections, a conservative estimate suggests that around 30 per cent, and possibly more, of the lime produced would meet the high quality requirements of the sugar industry set at a minimum of 60 per cent available lime (see Table 2 and Appendix 2).

Fuel consumption and thermal efficiency

The amount of fuel (charcoal) used was 34kg per hour, which is equivalent to 816kg per day. The thermal efficiency was calculated to be 36 per cent. A heat balance calculation indicated that less than 10 per cent of heat losses were in the exhaust gases and discharged lime. Although the heat of the discharged lime fluctuated in early trials during later trials it stabilized at around 150 to 200°C.

The exhaust gas heat losses are usually an important indicator of efficiency. During the trials, gas temperatures remained remarkably low at around 60°C.

There is, however, a discrepency of between 40 and 45 per cent. This constitutes heat losses through the kiln walls. In this instance, the difference is probably as high as it is are due heat loss through the cracks that appeared in the kiln and through the inspection holes.

6. BALAKA FORCED-AIR SYSTEM

The experience with the experimental forced draught kiln at Chenkumbi was very positive. The lime producers showed an immediate interest in it. One of the producers promptly built his own vertical shaft structure. We showed that suitable quality lime for the sugar industry could be produced by this kiln — based on the sugar corporation's own analyses. In addition, we were now in a position to make modifications to the kiln to improve its efficiency and its durability.

The next stage of the research and development effort was to develop a low cost hydration, milling and classification plant. We approached a local USAID funded enterprise development organization for financial support on this. They recognized the potential and agreed to assist.

A consortium of local business people was brought together to participate in the research and development phase with a view to taking over the project once this phase was complete as an operating and viable enterprise. A joint venture in this phase was, therefore, entered into with USAID providing financial and management input, and ITDG providing the necessary technical development skills.

The new modified forced draught kiln was constructed in Balaka town which is about 15km from the experimental kiln site at Chenkumbi Hills. The decision to locate the plant at Balaka was made jointly by the consortium and USAID on the grounds that it would be near to a power supply and water, and also near to transport infrastructure. These advantages were set off against the disadvantage of transporting limestone from a quarry at Chenkumbi Hills.

Quarrying

Limestone was quarried from a site at Chenkumbi Hills by means of first drilling, blasting and then manual breaking and dressing to kiln feed size with hammers. A quarryman with drilling and blasting skills would be hired every three to six months to extract sufficient rock for the period. The rock was then brought to the Balaka site by hired trucks.

Kiln design, operation and performance

The kiln was constructed with the same internal dimensions as the experimental kiln at Chenkumbi, i.e., an effective shaft height of six metres, 1.1 metre internal diameter tapering one metre from the top to 800mm diameter. The kiln was lined internally with locally produced refractory bricks. These were custom made for the first time in Malawi specifically for the emerging lime industry. The refractories were wedge-shaped fire-clay bricks, in two sizes, to facilitate the tapering of the shaft at the top. Outside the refractory lining was a 230mm layer of insulation bricks, once again made specifically for the purpose, and outside that was a 120mm skin of ordinary clay stock brick. The brickwork was encased in a 6mm mild steel casing with a 2.3m diameter.

The kiln is free standing and access to the top of the kiln was via a cat ladder built adjacent to four inspection ports. A steel platform was built on the top of the kiln, fitted with wooden decking, from which to charge the kiln.

Figure 5. Commercial kiln in Balaka, based on the
successful experimental kiln in the Chenkumbi Hills.

Kiln charging

The kiln charge was brought to the top of the kiln by means of buckets drawn up by a windlass which is fixed to a davit arm. The davit arm can swivel from over a charging hopper at the centre of the kiln to clear the kiln at the side. The charging hopper is a fabricated square sheet metal unit fitting over the round hole at the top of the kiln shaft. It has two flap doors fixed to counter weights to make sure that the kiln remains sealed except during charging. A 3 metre by 300mm diameter chimney leads from the side of the charging hopper. The limestone and charcoal was hoisted manually and charged into the kiln continuously.

Limestone charge:	22 charges in 24 hours 10 buckets by 26kg per bucket per charge 5720kg in 24 hours.
Charcoal:	22 charges in 24 hours 6 buckets by 6 kg per bucket per charge 792 kg in 24 hours
Labour:	4 people per shift x 4 shifts

Kiln operation

The temperature in the kiln was set at between 1000 and 1100°C The temperature was checked at the commissioning stage by means of thermocouples but later, once the operation stabilized, it was possible for the kiln operators to judge the temperature fairly accurately by visual observation of the colour.

Initially there was some difficulty in balancing airflow with retention time and discharge rate. While this kiln was built with the same dimensions as the kiln at Chenkumbi Hills, it did not perform in the same way. There could be a number of reasons for this. What is important to note is that this can be expected with each new kiln. Within four or five days from start up, however, it was possible to establish visually that the product from the kiln was being well calcined. There was no sign of charcoal in the discharge and a relatively low quantity of core (uncalcined limestone).

Forced-air unit

The kiln fan was installed with a horizontal bottom discharge, mounted on a concrete plinth above ground level, and driven by a 5.5kW motor. Air is fed into the kiln by the fan through a 150mm pipe to a series of cast-iron tuyères forming a manifold in the middle of the kiln.

The fan was calibrated using a pitotstatic tube to measure flow rates, and a manometer to measure pressure rise. The fan was set to give 650m^3/hr, to match the optimal flow rate observed during the Chenkumbi trials. At 650m^3/hr in Balaka the pressure drop was 20-25mm water gauge. This seems very low, but it correlates well with figures given by Wingate.[8] Although there is no comparative figure for the Chenkumbi Hills kiln the indications from the currents drawn are that the pressure drop there was higher (see Table 1). This might account for the initial difficulty in commissioning.

Kiln discharge

The lime is discharged from the kiln by means of shovels from four discharge openings directly into wheelbarrows which are used to carry the quicklime to the hydrator charging platform. A long crow bar was used to break any arches forming in the kiln and to thus drop the lime and limestone kiln mass. Metal fire doors were installed at the four discharge ports. The kiln discharge rate was set at every one hour and 30 minutes, compared with one hour 45 minutes at Chenkumbi. The kiln was provided with lighting at the charging platform for night time operation.

Input/output

The discharge frequency of one hour 30 minutes compared to one hour 45 minutes for the Chenkumbi kiln increases the output of the kiln and, therefore, also the amount charged into it. The amount charged into the kiln was 5720kg limestone. There was a loss on ignition of between 36 and 37 per cent, i.e., about 86 per cent conversion of the $CaCO_3$, resulting in a CaO product of around 3000kg per 24-hour day.

Approximately 970kg water is added to this in 24 hours to produce 4370kg lime hydrate, including waste. From this about 11 per cent is sieved off as plus 5mm waste which results in material to milling, classification and bagging of 3890kg. Sieve analysis of this material showed that over 70 per cent of the slaked and sieved hydrate was less than 150 micron.

Quality

The quality of lime produced averaged at around 60 per cent available lime, and the particle size distribution, after slaking and sieving, was shown to be 70 per cent less than 150 microns.

This quality is within the limits of quality required by the sugar corporation. The bulk of the imported material is of a quality well below this. The selling price, ex-works, for this quality is nearly double the price to the lower quality building market.

Fuel consumption and thermal efficiency

The fuel that can be used by this kiln is charcoal (hardwood or softwood), coal or fuelwood. Fuelwood was not tried, however, because its use is not considered a long-term prospect for lime producers. Hard and softwood charcoal, as well as coal were tried and all were found to be suitable. The choice, in effect, is between softwood charcoal and coal both of which come from the northern region of Malawi.

The fuel consumption was found to be 792kg charcoal or coal per day. Kiln thermal efficiency for the Balaka plant was 42 per cent compared to 36 per cent for the experimental kiln. The difference of around 6 per cent is due primarily to the insulation built into the walls of the Balaka kiln.

Experimental hydrator, milling and classification plant

Hydrator

It was decided to include a mechanical hydrator in the processing plant for two reasons: firstly, to reduce dust emanating from this activity, and thus improve the working environment, and secondly, to improve the quality of the product. Mechanical hydration makes for more even and thorough slaking, and a finer product is produced.

The hydrator designed and built for the plant is a batch operation machine comprising; a horizontal U-shaped mild steel trough two metres long and 800mm wide, a lid on top hinged on one side, a gate along the bottom for discharge, and a rotor internally consisting of a shaft and a series of angle sections welded to it alternately offset at 90 degrees.

Quicklime is charged into the hydrator at the top, and water is added to it through holes in the bottom of water pipes running along the top of the machine. The slaked lime is discharged through a gate running the length of the hydrator.

Hydrator trials and performance
The initial tests on the hydrator showed that it is very important for the hydrator drive to be started before any water is added. The effect of water addition prior to start up is an increase in torque required from the drive which is too high for the gearbox. After several attempts the trials resulted in a quicklime load, unsifted, straight from the kiln, of 220kg per batch with 60 litres of water. It is felt that the batch size could be increased to 300kg by reducing the rotor speed.

A number of different rates of water addition were tried. The trials were started with a rate of about 30 litres per minute. In these trials very little happened in the hydrator for the first 2 minutes, then the reaction became very vigorous and clouds of steam and dust poured from the vent and the top and bottom doors of the hydrator. The atmosphere around the hydrator became hot and choking during this period. The reaction then subsided.

The trials concluded at the rate of 10 litres per minute, at which the reaction was less vigorous. Trials to determine the amount of water required concluded at around 60 litres per batch which is equivalent to about 270 litres per tonne of quicklime. The time required to hydrate one batch was betwen 10 and 12 minutes. However, the overall batch time from starting one batch to starting the next was about 20 minutes. The hydrator requires four people to operate it per shift.

The product of hydration was hot and slightly moist on discharging but became dry within a few hours. It was also considerably more homogeneous than manually slaked lime hydrate.

During the early trials, the 100mm hydrator vent stack seemed to work well but later it blocked up. This was caused by steam condensing on the pipes and collecting lime dust. The vent was adjusted to allow cleaning, and this in turn created a serious dust problem. It was concluded that the size and position of the vent would need to be modified to deal with this problem. It was also found that the hydrator lid and discharge gate did not seal properly. Although emissions from these were greatly improved by adjustment to the machine on site it will need design modification to eliminate this problem completely.

The hydrator was specifically designed as a batch machine. The experience derived from commissioning and operating this machine has shown that a future model could be a continous flow machine. This would increase production rates and further reduce the dust problem.

Milling and classification plant
Mechanical hydration produces a greater proportion of finer and more evenly slaked lime which is within the specifications of the high quality market — shown by both physical and chemical analysis. These finer fractions need to be separated out from the coarser material which has a somewhat lower quality. The function of the processing plant is to do this and to mill the coarser fractions to required specification.

The plant was designed as an experimental plant. The behaviour of the lime produced in process, from hydration onwards, was unknown. The plant, therefore, had to cater for a wide range of possible behaviour. Consequently, the process flow was more complicated than would otherwise be the case.

At the start the plant consisted of an intake hopper, classifier, classifier fan, classifier filter, hammermill, cyclone and hammermill filter. The process flow was designed to be as follows:

○ Lime is fed into the intake hopper which feeds into a pipeline conveying the hydrate to the classifier. The classifier is a double whizzer type. It separates the hydrate into two size fractions, a product size which is drawn out of the top of the classifier to the classifier filter, and an oversize which passes down a chute to the hammermill.

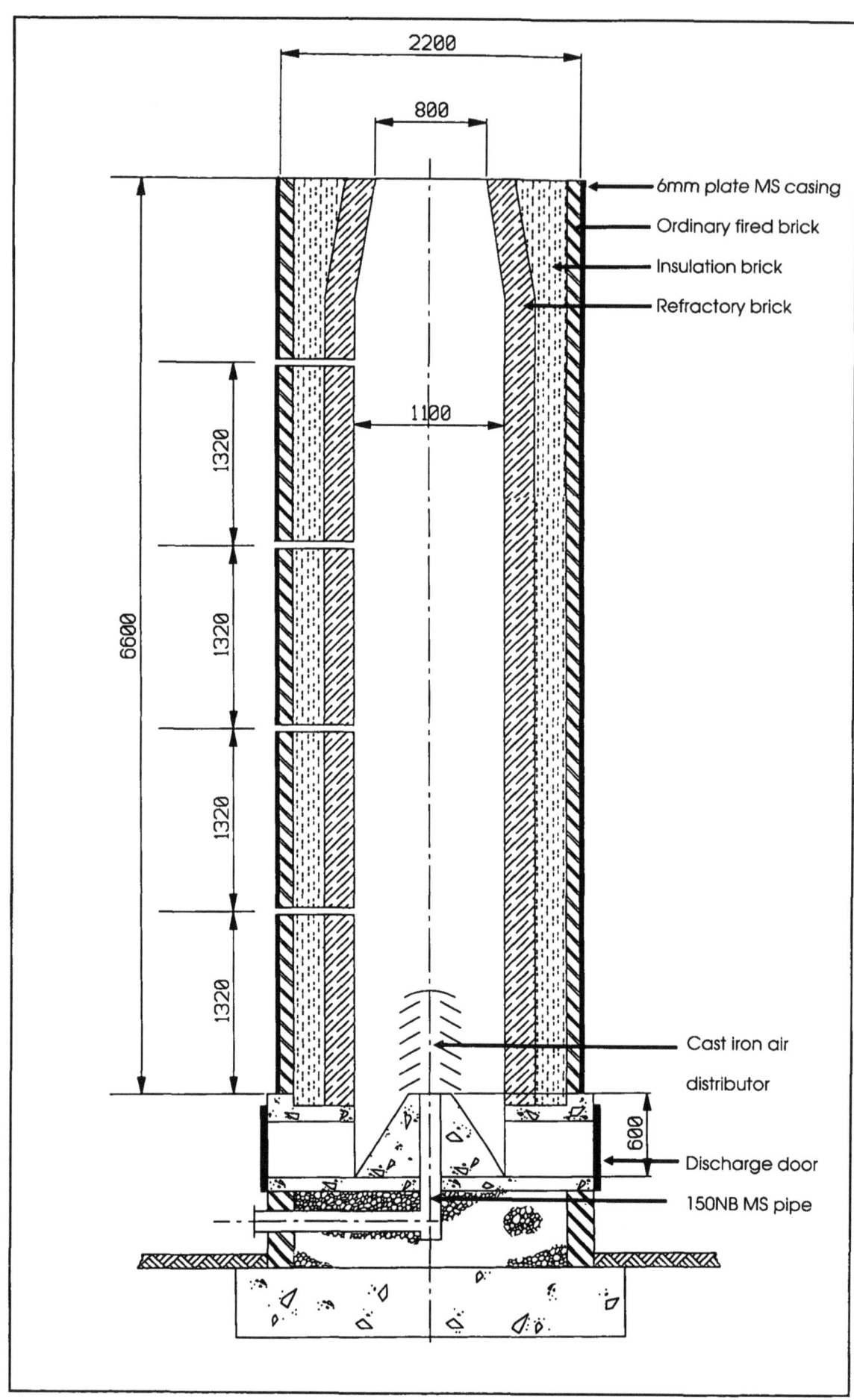

Figure 6. Technical detail of improved kiln design.

30

○ The material which is taken off by the classifier as product is a fine, high available lime product which can go straight to bagging. The oversize hydrate is milled in the hammermill to product size.

○ The hammermill is in closed circuit with a cyclone. The oversize is recycled back into the hammermill and the fines are carried through to the hammermill filter.

The operation of the classification and milling plant was fraught with problems from the outset. These problems were predominantly engineering in character. They caused substantial disruption in installation and commissioning which resulted in the budget for these activities being exceeded. Consequently the plant was inadequately commissioned. These problems have continued nearly a year after the initial commissioning. However, while they are near resolution, the experience deserves reporting on in a separate document and, therefore, will not be elaborated upon here.

APPENDIX 1

Uses of lime and lime hydrate

Metallurgical

Steelmaking flux; iron ore agglomeration; beneficiation of non-ferrous metals such as copper, alumina, and magnesium metal; and a wide range of other minor uses such as wire drawing and neutralization of metallurgical waste.

Chemical

Manufacture of sodium alkalis such as sodium carbonate, bicarbonate and hydroxide; inorganic salts and bases such as magnesium oxide and hydroxide; calcium carbide; and a host of other calcium chemicals such as hypochlorite, nitrate, soda-lime and precipitated calcium carbonate.

Sanitation

Water and effluent treatment

Ceramics

Glass manufacture, refractories and building products such as sand-lime bricks and cellular concrete products

Pulp and paper

Causticizing sodium carbonate solution.

Food and by-products

Sugar refining, leather tanning and citrus pulp processing.

Petroleum

Grease manufacture, petroleum refining and drilling muds.

Paints and coverage

Protective coatings, varnish, paints and pigments.

Construction

Masonry mortar, plaster and stucco, whitewash, lime cement paint and soil stabilization.

Agriculture

Direct soil liming; soil neutralization and fertilization; liming of forests, lakes and ponds (in industrialized areas).

The industrial applications referred to use lime in a variety of forms. Physically, it is used in sizes ranging from 0.05mm to 50mm. All industries apart from the construction industry and agriculture demand a lime of high chemical purity. Some industries however, use dolomitic lime, either in a soft-burnt reactive form or as a so-called dead-burnt dolomite. Another form used in hydrated lime ($Ca(OH)_2$), mostly in construction and agriculture, but also in sanitation and other industrial applications.

The processing required is calcination, crushing, grinding and sizing, if a quicklime product is required. Alternatively, the quicklime is crushed and slaked to produce a lime hydrate. Quicklime is marketed in granular or pulverized form. It is a material which is reactive in the presence of moisture. It is therefore stored and transported in airtight containers, to prevent air slaking and thus preserve the reactivity for which is is valued, and to make for safe handling.

APPENDIX 2

Experimental forced draught kiln trials

(Information extracted from Bush [2])

The first trial was not subjected to analysis as it was of short duration and the data was inadequate. The discharge frequency in Trial 1 had been set at 75 minutes and the air flow at 680m³/hr. During Trial 2, the firing zone dropped down the kiln shaft. At hour 13 (times quoted are from the start of Trial 1), the air flow was increased to 790m³/hr and the discharge frequency reduced to once every 90 minutes. The firing zone started to climb again but not as high as before. The available lime figure for this trial was 57/61 per cent. In any event, as it was a short trial this analysis should not be regarded as representative.

Trial 2 suggested both that 75 minutes was too short a period between discharges and that the air flow was set too high and as a result cooled the kiln. Between Trials 2 and 3 (Inter-Trial A and B), some experiments were carried out with air flows. The graphs on these trials show the response of the kiln to these changes. They are shown along the bottom edge of the graphs. They revealed that the kiln reacts rapidly to changes in air flow and that an air flow of around 650m³/hr is the optimum rate.

At 64 hours the fuel for the fan ran out. When it was restarted it was too high. This was not noticed at first so it was only readjusted after three hours. The kiln took about eight hours to settle down to a stable state. The stable condition was, however, far from satisfactory. The firing zone was far too low in the kiln. Red hot rock could be seen through the discharge ports.

Trial 4 was similar to Trial 3. A slightly higher air flow appears to have improved temperatures but the highest readings were still at the bottom of the kiln. A better quality lime was being produced but accurate separation was required to do this. A bar chart for Trial 4 shows particle size distribution and available lime analyses for specific size fraction of trial 4.

Following Trial 4 attempts were made to raise the firing zone. By discharging only at three hourly intervals for 9 hours it was possible to raise the calcination zone in the kiln to the desired position without changing the air flow. At this point it became obvious that the two major variables — air flow and discharge rate — were operating almost independently of one another. The air flow could be used to control the rate of combustion and hence temperatures, and the length of the firing zone, i.e., the period that the rock remained at calcination temperature. The rate of discharge governed the position of the firing zone.

Having established the desired temperature profile in the kiln, an attempt was made to stabilize it. Trial 5 shows the outcome. The air flow was initially set at 790m³/hr. This caused the firing zone to become compressed with the bottom temperatures dropping steadily. The air flow was adjusted to 650m³/hr and a stable position resulted with the firing zone centre at about the middle of the inspection hole. Before this trial, it was also decided to reduce the amount of fuel to increase efficiency. Trial 5b began to show good results — good available lime figures and high efficiency.

Given the outcome of Trial 5b, it was decided to decrease the discharge frequency further to improve lime quality. Trial 6a showed the best result of all. At hour 143, the diesel engine was misadjusted (6b) and at hour 150 the fuel changed to 50:50 mixture of wood and charcoal.

TABLES

1: Forced draught experimentation results

Trial number	Charge rate (kg/hr)	Discharge frequency (hrs)	Stone:fuel ratio	Fuel (kg/hr)	Air (mm)	Air flow (m³/hr)
1	233	1.00	5.9	40	32	700
2	209	1.50	5.9	35	31	800
3	205	1.50	5.9	35	18	650
4	253	1.50	5.9	43	25	740
5a	326	1.50	7.1	36	32	810
5b	275	1.50	7.1	39	18	650
5c	275	1.50	7.1	39	18	650
6a	241	1.75	7.1	34	18	650
6b	245	1.75	5.3	47	18	650

Stoic air (m³/hr)	% excess air	Available Lime			F_c[1]	Kiln efficiency[2] (%)
		mixed	cores	fines		
382	83	48			0.62	22
342	134		57	61	0.77	27
342	90	47			0.61	22
412	79	54			0.70	25
440	84	45			0.58	25
371	75		58	60	0.76	34
371	75		57	59	0.75	33
325	100		59	66	0.82	36
322	102		47	66	0.75	33[3]

NOTES

1. F_c is the fraction of calcium carbonate ($CaCO_3$) that has been calcined in the kiln. Its calculation is based on 10 per cent unburnt cores and 40:60 core:fines ratio.

2. Kiln efficiency is the proportion of heat (calorific value) that has been used for calcination. The calculation assumes a heat of calcination is 2740KJ/kg. The figures show the fractional efficiency, i.e., they are corrected for the fraction of the calcium carbonate converted to lime.

3. Efficiency of trial 6b based on wood/charcoal mix.

4. Figures for available lime content in Table 1 are analyses of bulk samples of the 'fines' and 'core' material and not analyses of specific size fractions.

2: Fuel consumption and efficiency

FUEL CONSUMPTION AND EFFICIENCY	Traditional kiln	Natural draught kiln	Forced draught kiln — Chenkumbi	Forced draught kiln — Balaka
Limestone charge per day	75 t/batch	7200kg	5500kg	5500kg
Fuel charge per day	55 t/batch	2880kg	816kg	792kg
CaO produced	42 t/batch	4280kg	3000kg	3400kg
Fraction of CaO$_3$ converted to CaO	0,58	0,68	0,82	0,86
Fuel used per tonne CaO produced (MJ)	19640	10090	8160	6928
Calorific value of fuel	15MJ/kg	15MJ/kg	30MJ/kg	30MJ/kg
Fuel cost per tonne lime produced (MK)	94,00	54,00	40,80	34,95
Kiln efficiency (%) assuming 100% conversion —	15	29	36	42
assuming fractional conversion —	9	18	29	36

* CaO content of limestone 91.4 per cent.

Kiln or thermal efficiency is the ratio of heat used in calcination to total heat consumed. It ranges from 15 per cent for the simplest wood fired batch kiln to 85 per cent for the most efficient modern kilns. Efficiency is calculated using the following formula:

$$\% \text{ kiln efficiency} = \frac{H_c \times L_s}{C_f \times M_f}$$

where:
H_c is the heat of calcination per tonne CaO produced (theoretical heat required). This is 3200 MJ/tonne CaO.
L_s is the available lime content of limestone
C_f is the calorific value of the fuel.
M_f is the amount of fuel per tonne CaO produced, in kg.

The formula assumes 100 per cent conversion of the limestone to lime. A more accurate reflection of efficiency would be fractional conversion, i.e., full conversion kiln efficiency multiplied by 'Fraction of CaCO$_3$ converted to CaO' factor (see above).

3: Comparative capital costs

CAPITAL COSTS (All figures in Malawi kwacha)*	Traditional box kiln	Natural draught vertical shaft kiln	Forced draught[3] vertical shaft kiln
Kiln	500	10000[1]	59000[4]
Hydration, classification and milling	–	70000	229500
Buildings	4500	6500	90500
Sub-total	5000	86500	379000
Working Capital	10000	26620[2]	49150
TOTAL INVESTMENT	15000	113120	428150
ANNUAL RETURN	1425	21420	58820
ANNUAL RETURN ON INVESTMENT	9.5%	18.9%	13.7%
Depreciation cost (per annum)[5]		5600	23170
Loan finance requirement (assumed)		90500	342500

NOTES:
1. Cost of kiln includes all materials — steel straps, chimney hood, ramp and kiln. Kiln lining is traditionally produced from fired brick which would be totally replaced every six weeks (one week to replace). It also includes labour and cost of supervision. No amount included for commissioning or training.
2. Two months' production cost.
3. Based on revised plant cost estimates.
4. Kiln including installation and commissioning
5. Plant and equipment depreciated on a straight line basis with no salvage value over 15 years, and buildings depreciated in the same way over 25 years.

* Five Malawi kwacha is equivalent to £1 sterling (approximately).

4: Comparative incomes statement

INCOME STATEMENT (All figures in Malawi kwacha)*	Traditional production	Natural draught production[8]	Forced draught production[9]
Sales (2000 25kg pockets lime)	10000	10000	17000[6]
Production costs:			
Labour cost[2]	930	1005	1640
Blasting cost	–	475	470
Fuel cost	3960	2,325	1535
Bags (2000 pockets @ 50t/pkt)	1000	1000	1000
Maintenance costs:			
Production	3000[4]	1180	1880
Supervision	180	300	500
Total production costs (per 2000 pockets)	9070	6285	7025
Overhead costs	455[1]	940[3]	1055[3]
Loan repayment[5]	–	1440	5375
Depreciation costs	–	270	1930
Total costs per 2000 pockets lime	9525	8935	15385
Gross Profit per 2000 pockets lime	475	1065	1615
Production per annum (tonnes Ca(OH)$_2$)	150	1050	1065

NOTES:

1. Overheads calculated on basis of production of three batches per year (5 per cent of production costs).

2. Daily government minimum wage rate used in calculation.

3. Overhead set at 15 per cent because of continuous production.

4. Milling cost charged at MK1.50 per packet.

5. Repaid over five years in equal installments. The interest rate is 20 per cent (estimate). Calculated using @PMT function on Lotus123 spreadsheet programme.

6. 600 bags at MK5.00/packet and 1400 bags at MK10.00/packet.

7. Charcoal cost per tonne is MK150.00 delivered to site.

8. 210 working days. Due to additional repair and maintenance.

9. 275 working days.

5: Comparative economic analysis

ECONOMIC ANALYSIS	Traditional production	Natural draught production[1]	Forced draught production[2]
Production per annum per production unit	CaO: 42-210 Ca(OH)$_2$: 240	CaO: 898 Ca(OH)$_2$: 1,050	CaO: 935 Ca(OH)$_2$: 1,065
Total workplaces in industry[3]	Perm. 85 Casual 965[4]	Perm. 350	Perm. 400
Capital cost per job created (MK)	1250[5]	2260	7,510
Number of production units[3]	29	7	7
Foreign exchange savings per annum	–	–	MK1.26 million UK equiv.: £252 000
Annual cost per workplace (MK) (includes casual labour)	77.05	422.00	572.90

NOTES:
1. 210 days production per annum.
2. 275 days production per annum.
3. Assuming production meets demand. Total demand estimated to be 7,000 tonnes hydrate per annum.
4. Majority of workers (approx. 75 per cent) are casual labour paid piece rates.
5. Excludes cost of milling plant.

References

1. Boynton, R.S. *Chemistry and Technology of Lime and Limestone.* 1980. John Wiley and Sons, Inc. New York.
2. Bush, A. 'Balaka Lime Kiln: Report of trials conducted in April 1989'. May 1989. (ITDG report)
3. Jones, B. (Aptech Zimbabwe). 'Balaka Lime Project: Final installation and commissioning report'. December 1990. (ITDG report)
4. MacAllister, P. 'Proposals for upgrading the small-scale lime producers in Malawi'. June 1988. (ITDG report)
5. Spiropoulos, J. 'Mini-lime Plant: Technology profiles'. 1986
6. Spiropoulos, J. 'Balaka Lime Project: Report and draft project proposal'. July 1987.
7. Whitten D.G.A. and Brooks J.R.V. *The Penguin Dictionary of Geology.* 1981.
8. Wingate, M. *Small-scale lime burning: A practical introduction.* IT Publications 1988.

9 781853 391446